# CHICKEN POOP

The Subtle Art of Raising Backyard Chickens

EARTH HARMONY LIVING

Chicken Poop

The Subtle Art of Raising Backyard Chickens

By Guenevere MacDonald

Editor Bethany Rickard-Wood

Photography Guenevere MacDonald, Andrea Westover

Published by Earth Harmony Living

Copyright 2020 BY Guenevere MacDonald All rights reserved.

In no way is it legal to reproduce, duplicate, or transmit any part of this document in either electronic means or in printed format. Recording of this publication is strictly prohibited and any storage of this document is not allowed unless with written permissions from the publisher. All rights reserved.

The information provided herein is stated to be truthful and consistent, in that any liability, in terms of inattention or otherwise, by any usage or abuse of any policies, processes, or directions contained within is the solitary and utter responsibility of the recipient reader. Under no circumstances will any legal responsibility or blame be held against the publisher for any reparation, damages, or monetary loss due to the information herein, either directly or indirectly. Respective authors own all copyrights not held by the publisher.

Legal Notice:

This book is copyright protected. This is only for personal use. You cannot amend, distribute, sell, use, quote or paraphrase any part or the content within this book without the consent of the author or copyright owner. Legal action will be pursued if this is breached.

Disclaimer Notice:

Please note the information contained within this document is for educational and entertainment purposes only. Every attempt has been made to provide accurate, up to date and reliable complete information. No warranties of any kind are expressed or implied. Readers acknowledge that the author is not engaging in the rendering of legal, financial, medical, or professional advice.

By reading this document, the reader agrees that under no circumstances are we responsible for any losses, direct or indirect, which are incurred as a result of the use of information contained within this document, including, but not limited to, errors omissions, or inaccuracies.

1. Edition, 2020
© All rights reserved.

Earth Harmony Living

ISBN 9798552353125

*Dedicated to My husband, who has been my fellow Chicken keeper and loving partner in crime for all these long years.*

*To my daughters that they will continue to love and be inspired by nature and all its creatures.*

*And my wonderful support and editorial team Bethany Rickard-Wood and Andrea Westover for supporting and putting up with me.*

.

# Table of Contents

Preface

Chapter 1 What Do Chickens Need

Chapter 2 Set Up

Chapter 3 Why not to Keep Chickens.

Chapter 4 Which Birds

Chapter 5 Getting Started with Vocabulary

Chapter 6 Getting Started with Chicks

Chapter 7 Getting Started with Laying Hens

Chapter 8 What not to do with your Birds

Chapter 9 Cleaning up after your Birds

Chapter 10 Winter Basics

Chapter 11 Dealing with Predators

Chapter 12 Know before they Crow

Chapter 13 Daily Task List

Chapter 15 Chicken Diseases

Epilogue

*"There are tiny dinosaurs in the barn."*

*"That's where they are supposed to be."*

*Melodie & Teagan*

## *Preface*

Every good chicken book starts with the question, "Why keep Chickens?".

The answer tends always to be pretty much the same list every time.

Farm Fresh eggs for starters. Who would not love firmer, deeply colored, more nutritious eggs? Eggs are probably the number one benefit of having backyard chickens. From the different colors laid by different breeds to the different sizes and laying patterns of the hens, eggs are a much anticipated and celebrated part of chicken rearing.

Then there is the hormone-free meat. Since you control what goes into your birds, you control what comes out of them too. You control the meat process from start to finish. By raising your meat, you also help reduce the carbon footprint left by grocery store deliveries.

Organic compost and fertilizer. Like your meat and eggs, you control what goes into your compost. Chicken poop is an excellent organic fertilizer and contains high nitrogen levels, suitable for the garden and house plants. Just remember that chicken poop is a hot compost meaning it must compost for a minimum of 6 months (preferably longer) before it can be applied to gardens. Mixing your chicken waste with other organic materials will ensure a rich nutrient-based compost that will make your garden flourish.

Pest control. Ahh, how chickens love bugs, from beetles to ticks to fleas, they will eat all the bugs they can catch. They will also catch and eat mice, voles, moles and frogs and toads, and in some instance's small snakes.

And of course, this is where amusement is also added to the list. Chickens are fun and relaxing to watch and often get themselves in some stupid circumstances, making pure entertainment fodder.

It is pretty much an industry standard list. But there are some things that are not found on regular run-of-the-mill lists. For starters, chickens are the closest living relative of the tyrannosaurus rex. Who would not want a bunch of tiny dinosaurs roaming the backyard?

Chickens also have their own social network... they were tweeting before Twitter came on the scene. They have their own language, which is quite extensive, and they start talking in the egg. They even purr when they are happy.

Chicken eyes work independently of each other, and they can watch you and their food simultaneously. They also see movement 20x faster than humans, so they can spot predators faster and make a run for safety. Chickens are pretty friendly animals that can be trained to come when their name is called. They can be trained to do tricks as well. And while there is work involved, raising chickens is a relatively easy affair. A small coop can be kept in a backyard in the city, allowing for 3 to 12 birds depending on ordinances. In the country, you can have larger flocks free-ranging your front yard. Nothing is quite as enjoyable as watching chickens at play in the yard. There really is not a good reason not to keep chickens if you can. So how do you get started? I am so happy you asked...

## Chapter 1 What do Chickens need?

***Space*** - a typical quarter lot can keep as many as 12 chickens without issues. If they are free-ranging only. 4ft of space for the average medium bird in a coop and 8 to 10 ft outside if penned. These are the recommended numbers; however, depending on your chickens' size and needs, 2 ft in the coop may be sufficient. Odds are they won't even use their full 2ft of space, but having it there is good. Outside birds tend to explore their surroundings while staying close to safety. After they have eaten and explored, they usually come back closer to their coop or the door of the run to sit and observe things. Having sufficient space allows them to fill their curiosity and stretch and dust.

***Ventilation.*** I cannot stress this point enough. Ventilation is necessary year-round. Every coop should have at least 1 to 2 inches along the ceiling where the roof meets the ceiling that is fenced in but open to air circulation. Having ventilation above the bird's heads allows for ammonia and heat to rise and escape in the summer and for moisture and ammonia to escape in the winter, preventing frost bite and respiratory issues. Often, it's thought that birds need heat and be sealed in completely, but that actually is the worst possible thing that can happen in a coop, especially when it's really hot or cold. Ventilation is an absolute must for chicken health and wellbeing.

***Protection from predators.*** Oh, the predators... snakes that eat eggs, raccoons, coyotes, foxes, weasels, minks, hawks, and owls. And then there are the domesticated predators known as the neighbor's dog. There are plenty of things out there that love a fresh dish of chicken, and they can wreak havoc, killing an entire flock in one go just trying to get one solitary bird. All coops and runs should be predator-proof. All ventilation shafts should be secured with Fencing. Fencing should be hardwired cloth or other heavy-duty Fencing. Chicken wire is a no-go. It was invented merely to keep chickens out of gardens, not predators out of coops and runs.

Remember, there as well are two ways to fence.

Fencing in keeps your chickens inside the run with the Fencing on the inside of the post., Meaning they cannot escape by pushing on the wire because the posts stop the fence from falling. The same goes for fencing out. This is the term for keeping predators out of the pen. When you fence out, the Fencing is on the outside of the poles on the predator side. Again, if the animal were to push on the fence, the posts would stop it from coming down. All coops and runs should be fenced out against predators.

Fencing should also be buried a foot down and or out from the fence to keep predators from trying to dig under the fence. If you have a run, you should also have overhead protection from Hawks and owls. This can be done with shade tarps, Fencing, fishing line, and vining plants (make sure they aren't poisonous to chickens first). The coop should also be predator-proof, with all small holes and gaps along the ground secured and ventilation fenced accordingly.

***Fresh water and proper feed.*** Birds must have fresh water daily and extra water in the hot summer months as they are prone to heatstroke. The feed should be a good quality layer formula for laying hens and roosters. Pullets and chicks should have a balanced starter or grower formula. Please note that chicken scratch is not feed. It is an extra treat for birds but does not have the nutritional content needed for your birds to be healthy.

***Nesting boxes*** (1 box per 4 laying hens). Nesting boxes can be anything from a dresser with a couple of drawers removed. Extra drawers on the side. Kitty litter containers on their sides, baskets, and commercial nest boxes. Birds do not care so long as they have a safe, comfortable place to nest. You can get highly creative with nesting boxes.

***Wind and element protection.*** Just like us, birds do not care to be standing around in the wet and cold. Whether it be an open coop, foliage, or chicken teepee shelter, having a place is essential to keeping healthy chickens. While some might not mind a bit of wind and rain (even snow), they all eventually want to get out of the elements to dry off and warm up, so shelter is essential.

***Heat*** (not always necessary). Heat in a coop is a touchy subject. I am a firm believer that heat is not needed in a chicken coop any more than in a horse or cow barn. If your coop is the right size for the number of birds, they will use their body heat to warm it up. Birds will fluff up their wings and trap warm air under them. They then fluff up and cover their feet. By having roosts up off the ground, they also take advantage of the warmer air rising, and tricks like deep littering methods can produce additional heat in the coop without the threat of fire. A heating lamp in the coop runs the risk of fire; from dust, bedding, and feathers. There are too many things that can catch on fire from a heat lamp. The alternative is lighting that is covered. We use led barn lamps in our coops and barn. These are covered and enclosed, so the bulbs never come in contact with fire hazards like feathers. They do not produce heat (per se), but they provide additional light for the birds. We live in a climate that reaches -49 C in the winter, and our birds have been fine in an unheated coop.

***Bedding.*** So many choices. Our preference is Pine Shavings (do not use cedar). They absorb well, are compostable, and relatively inexpensive. We use the pine shavings for both our chickens and our rabbit pens. Straw can also be used, but I do not recommend it as it is hollow and has a smaller surface for absorbing moisture. Straw also tends to mold quite easily, causing respiratory issues. Hay is never recommended as it does not absorb anything and would need to be replaced daily. Straw every other day, and pine shavings can go weeks or even months if you use the deep litter method. Sand is another option that is highly encouraged, along with pine shavings. The only downside I have seen to sand is the increased risk of bugs such as fleas. However, chickens love to eat bugs, so the two can sometimes sort themselves out.

## Chapter 2 Set Up

Where to find your birds. Depending on what you want from your birds will determine where you get your birds. If you are looking for specific breeds, then hatcheries are an excellent source for chicks. Local breeders will have hardy birds in your area, and local feed stores often have chicks and can order in laying age hens. Even the SPCA can be a source for laying hens or roosters as they are sometimes abandoned or found running loose.

Laying hens or chicks. If you want eggs right away, then laying hens are your best bet. These are birds between 16 weeks and two years that are laying regularly. After two years, their egg production slows, and they don't lay as much. Many chicken owners chose dual-purpose birds for this reason so they can be meat when they no longer lay regularly. Chicks are usually sold at a few days of age and must be 16 to 20 weeks before they lay eggs. Chicks require more care than hens and must have separate housing from older birds.

Housing for birds can be anything from a portable chicken tractor, a standstill chicken coop, or a chicken pen built in a barn. Suitable chicken housing should be predator safe, have proper ventilation and sufficient space for the number of birds housed, and potential future birds such as chicks the need to grow up.

Feeders, waterers etc. Feeders and waterers come in different styles from open pan bowls to hanging feeders that allow free feed. Waterers should be secure so they don't leak into the bedding and shouldn't be so large that young chicks can get trapped in them and drown.

Space - The standard rule of thumb is 2 to 3ft of space in the coop for a medium size bird and 8 to 10 feet of outdoor space if you are not free ranging.

Fencing vs free range - If you decide to have chickens you are going to have to decide whether you will fence your birds in or permit them to free range on the property. Both have pros and cons from predators to bug control. Ultimately the amount of space you have for them to roam will also play a huge part in that decision as well as known predators in the area.

Nesting and bedding supplies - Nesting and bedding supplies should be a material that is highly absorbent, inexpensive, and easy to clean. Pine Shavings and sand are the best options for bedding supplies. While shavings and straw top the list for nesting boxes.

## Chapter 3 Why Not to Keep Chickens

Like everything, there are downsides and cons to raising chickens in your backyard. Number one is where you live. If you are close to neighbors, they may not care for chickens wandering into their yard or a rooster crowing all day. City ordinances also play a vital role. In most instances where chickens are permitted within city limits, the number of birds is usually restricted to 5 or less, and roosters are forbidden. So, keeping secret birds in the backyard may be a no-go if you get caught. Don't assume either that because another neighbor has chickens that you are permitted to as well. Always check with your city or town council for the rules.

Noise is another downside to chicken raising. While roosters are slightly less loud than a barking dog. Roosters cannot be told to be quiet. They cannot be taught to be quiet either. And contrary to popular belief, they do not crow early morning when the sun rises. They can crow overnight and early morning, and all day long if it suits them. On the other hand, Hens are incredibly quiet in small numbers, so many cities will permit hens but not roosters.

Dust. Chickens stir up dust like nobody's business, in the yard in the coop in the run. Nothing makes them happier than rolling in dust and dirt. Unfortunately, this also means digging and scratching where they do not belong.

Manure. Well, it is a given everybody poops, and chickens are no different. Depending on your bedding choice, this can be an easy chore or an aggravating one, but poop will happen, and it will happen everywhere your birds go. Chickens do not have a sphincter muscle, so they have zero control over their waste and cannot, like other animals, be toilet trained. So, where you house your chickens and let them get their exercise, run, pen, yard, etc., expect to find chicken poop. But do not be discouraged by it. Chicken poop is an excellent compost material and can do wonders for your garden and house plants. Just be sure to compost it down as its nitrogen levels are too high when fresh to apply to plants.

Scratching is inevitable and can be very destructive. Chicken wire was invented to keep chickens out of gardens and flower beds and unless you have no issues with your birds scratching up all your flowers and vegetable, then penning them or fencing them out of the gardens with chicken wire is your best bet. On the flip side, this can also be a pro to having chickens as letting them loose in your garden at the end of the season to scratch and dig can significantly benefit the soil.

Allergies. This one is pretty much a no-brainer. If you are allergic to birds, then chances are chickens are not for you because sooner or later, you are going to have to clean out that chicken coop (or bribe someone else to do it.). If you have an allergy to the birds themselves, then picking them up is a no go and you can maybe work around it. But if its feathers do not even try. Feathers are everywhere; from pecking to molting to mating, feathers regularly fly throughout the coop the run the yard. They are everywhere.

Butchering is not for anyone who becomes attached to their birds. Rule of thumb if you plan to eat it, do not name it. We do not do the butchering of our birds on our property if we can send them to a local butcher. We do this for several reasons. Although we can and, if necessary, will butcher a bird on-site, we are also keenly aware of the predators in our area and chose not to draw them in with a butchering day. Butchering is not for the queasy animal lover either. It's a messy business that can take all day, depending on the number of birds. If you think butchering is not something you want to bother with, then be sure there is a butcher in your region that does chickens. Our local abattoir charges $3.50 a bird to butcher live birds—others in the area charge from $3.50 to $ 6.00 a bird. Turkeys tend to cost more at $8 to $10 a turkey or larger duck. If you have many birds you plan to butcher for meat; this expense should be considered upfront before raising the birds. Slaughtering on your own also requires special equipment such as killing cones, pluckers, and heat and cold tanks for the carcasses. These can often be rented depending on your region, although if you plan to raise and butcher large numbers of birds yearly, you should probably calculate the rental or purchase price in your startup costs ahead of time.

While there are several reasons not to raise backyard chickens, most of the cons can easily be worked around, making it feasible to do so. Some cons, such as allergies and ordinances, are not workable. In the long run, the pros to raising chickens outweigh the list of cons. Backyard chickens are a good investment for a healthy source of eggs and meat as well as their many other benefits.

*Chapter 4 Which birds*

Deciding which birds to purchase depends on what you want them for. Are you looking for meat or eggs, or both? Do you live in a hot climate or a colder one? Do you need a lot of eggs or just a few per week? Do you want a colorful flock with rainbow eggs, or is a monotone flock with the same-colored eggs simply fine? The amount size and color of eggs depends on the breed of bird. Meat quantities are a serious consideration as well. Leghorn chickens, for example, are laying machines laying up to five eggs a week per bird. However, they are small birds and do not produce much meat if they are worthy of a good soup. A meat bird like the Cornish cross is raised for 9 to 12 weeks then butchered for meat. (roughly 4lbs and up). However, Cornish crosses do not live long enough to produce eggs in most instances, and those that do are usually infertile. A Plymouth rock on the hand is a larger size bird that will lay up to 3 to 4 times a week and produces an excellent plump carcass for meat when she's done. Plymouths are great foragers while leghorns are not, and Cornish crosses often lose the ability to walk before butchering and can only go very short distances due to weight factors. The purpose of your bird will determine the type you want and need.

**For Eggs**. If you wish to raise birds for egg production, only the following birds are excellent for this purpose.

**Leghorn**–the powerhouse of egg-laying. The most prolific laying breed. The egg industry uses these small birds to produce eggs for the grocery chain. White with red combs and wattles, they lay medium to large white eggs 5 to 6 times a week. Red Sex Links–a medium-sized red-orange bird also used by commercial egg farmers. These birds lay brown eggs on average four times a week and produce a small carcass of meat when butchered. Easy going, they are easy to keep, good foragers, and do well in cold climates.

**Rhode Island Red**–a dual purpose bird Reds are good layers producing 3 to 4 brown eggs weekly. They can be a bit standoffish, and roos, although particularly good with the flock, can sometimes have an attitude with humans.

Australorps–large black birds. These birds are great dual-purpose birds that produce 3 to 4 brown eggs a week and have a good meat carcass of roughly 4lbs and higher when butchered. They are beautiful birds that do well in warm and cold climates. They are not prone to be flighty or jittery.

**Plymouth Barred Rock**- an excellent layer and mother, these birds lay 3 to 4 brown eggs a week. They make great dual-purpose birds for eggs and meat and are a colorful addition to any flock with their black and white barred stripes. Roos tend to be lighter colored as chicks and develop a greyer plumage while the hens stay dark. Easy to raise, good with people they do well in warm and cold climates.

**Orpington**- available in several colors, the Buff Orpington is the most common coloration. Prolific layers laying four eggs a week they are sweet and friendly and can be easily trained to come when called and jump for treats. They are an excellent dual-purpose bird, amazing mothers, and fantastic foragers. They do well in both hot and cold climates.

Colored Eggs. If you are looking for colored eggs, the following birds are known for their variety of colors but do not make good dual-purpose birds as they are smaller in stature than most laying breeds.

**Ameraucana**–carrier of the blue gene they lay medium blue to green eggs three times a week. Small and compact, they do well in all kinds of housing setups and are good foragers. They do well in both hot and cold climates

**Easter-eggers**–the mutt non-pure Ameraucana birds. These are birds that are mixed with other breeds but still carry the blue gene. They can lay eggs of a variety of colors depending on the mix of their breeding. Colors can range from blue to green, violet, olive, etc.

**Crested Cream Legbar**–barred blue egg layers. Legbars lay sky blue colored eggs 2 to 3 times a week.

They are friendly and sociable and do well in all climates.

**Olive egger**–These are Ameraucanas crossed with brown egg layers to produce generations of olive-colored eggs. Commonly Ameraucanas are crossed with Plymouths or Marans to achieve this coloration,, although any brown laying breed crossed with the blue gene will have the Olive coloration.

**Marans**- These come in different styles French Marans, Blue Marans, Coco Marans, Black Copper Marans. All of which lay dark brown, almost chocolate-colored eggs 2 to 3 times a week. The birds are larger than most colored egg-laying birds and can be used as dual-purpose birds. They are great foragers, friendly and pleasant,. Roosters are excellent flock protectors and friendly with humans. They do well in all climates.

Other breeds for egg laying.

Sussex, Wyandette, New Hampshire, Barnvelder, Hamburg and Lohan brown.

Other breeds for Colored eggs (pink, peach layers)

Light Sussex, Javas, Silkies, Orpingtons, Favorolles

Dual Purpose birds

Sussex, Delaware, Rhode Island Red, Plymouth Barred Rock, Red Rangers, Orpington, Australop, New Hampshire

Keep in mind that these are the most common breeds for these purposes, and several other breeds can also be included for dual purpose and egg-laying. Crossbreeds are also a good choice for raising dual-purpose egg layers.

How many eggs? If you are getting birds specifically for fresh farm eggs, then there are several things to consider. How many people are in your family? How often and how many eggs will you use? Do you plan to sell surplus eggs to friends or family? On average, a family of four with a dual-purpose bird such as the barred rock would need roughly 4 Plymouth barred rock birds to have roughly 16 eggs a week. So just a little over a dozen eggs. Suppose you use twice that much than you would need twice as many birds. If you use fewer than 4 Ameraucanas could bring in roughly a dozen eggs a week. If you want to have extra for baking and selling, then a leghorn chicken could produce 5 to 6 eggs a week, sometimes 7, so four leghorns would bring in roughly 24 eggs a week, so a dozen for the family and a dozen to sell. If you wish to have several different varieties, be aware of their laying average and add them together. Depending on space and ordinances, you may only be able to have the number of birds deemed permissive for your family, so chose your breeds wisely.

How much space do you have for housing? Space is another primary consideration for keeping chickens. Chickens can produce enough heat in the winter to warm themselves and their coop if the correct number of birds are present. Likewise, they can stay cooler in the summer if there is adequate space to keep birds spaced out enough to cool themselves. The same goes for enclosure space and free-range space. The average medium size bird needs about 3ft of space inside and 8 to 10 feet outside. There is some wiggle room on this; for example, if your birds are only inside for sleeping, then 2 ft of space is enough for one bird to sleep.

However, if they will spend time inside all winter and during inclement weather the rest of the year, then a minimum of 3ft should be provided. Overcrowding can mean overheating in the summer and respiratory issues year-round.

Do you want birds for show? Yes, people show chickens, and there are some exotic-looking breeds out there. If you plan to show birds, then coop size and condition are significant factors to keep birds from becoming ill and overcrowded. For show birds, egg numbers tend to be less than the other birds.

Do you intend to keep your birds over winter? If you want to keep chickens over winter, you want to pay very close attention to the space per bird. While insulation and heat lamps are unnecessary, space will play a large part in how warm or cold the coop is, even in the colder climates. By assuring that your birds have adequate space (not too much, not too little), the birds themselves can warm themselves and others without the intervention of heat lamps and heaters. Birds will fluff up their feathers and wings to trap warm air and cover their feet on the roost. They will huddle together for added warmth and the heat off the bedding rising also provides a source of warm air. If you want additional warmth, then the deep litter method (we cover this further on) will help as well. Size of your birds and their space are the most important criteria. If you are unable to house chickens over the winter months, then dual purpose birds that can provide eggs through spring and summer and a chicken carcass for the fall/wintertime frame would be ideal. If you plan to overwinter birds check your breeds to ensure they are birds that do well in colder climates as well. Birds with large combs and wattles tend to tolerate cold less than birds with smaller or nonexistent combs and wattles.

## Chapter 5 Chicken Vocabulary

**Chick**- a chicken under the age of 8 weeks

**Pullet**- hen eight weeks of age to egg-laying maturity (16 to 20 weeks)

**Hen**- laying female chicken

**Cockerel**–an immature rooster under 20 weeks of age

**Cock or Rooster**- a mature male of breeding age roughly 16 to 20 weeks plus depending on the breed

**Vent**- the part of the chicken that delivers both eggs and compost

**Crop**- a bag-like part of the chicken's digestive system visible in the neck and upper chest area on one side when full. Usually full by mid-day and will empty overnight as feed passes through the digestive system.

**Beak**- chicken mouth nose and defense tool. (not to be confused with a bill which is for waterfowl and geese)

**Spur**- found on some hens but mostly on roosters... used for defense. Located in the "ankle" region just above the foot.

**Comb**- found on some hens but mostly on roosters Larger on roosters than hens in most breeds. It is a band of soft tissue that can be in different formats based on breed. Example Rose comb, pea comb etc.

**Wattles** - Prominent on some breeds for both hens and roos, mostly prominent on roos.

**Saddle feathers**- locate just in front of the tail region... rounded on females, long and arrowed on males

**Bantams**–mini chickens... perfectly cute and useless

**Standard**–Regular breed size

**Large**–usually a dual-purpose breed.

*Chapter 6 Getting started with chicks*

One day old chicks can be bought from a supplier or from a hatchery and from local breeders. Many feed stores order chicks in the spring to sell on-site or permit you to request specific numbers of your breed's choice. These usually come from hatcheries, and the feed store works as a middleman in the purchase process. Day-old chicks can travel for several days with only the yolk they absorbed in the egg for food. However, once they arrive, they should be given water and an appropriate chick starter feed.

Chicks do not require fancy housing; however, they do need heat. Here on the farm, we start our chicks out in a Rubbermaid-style bin with feed and water and a hanging heat lamp. We pay special attention to the type of bulb (we never get tinted bulbs as they give off fumes), and we keep the light much higher than the bin. We also have heating pads for the chicks to stand on when we leave the house and do not want to leave the heat lamp unattended. All our chicks are raised in the home until they have feathered out sufficiently and the weather is adequate to allow them out. They go to their own pen in the barn or depending on the time of year to a chicken tractor of their own from the house.

Young chicks mustn't be in with older chickens unless they have a mother hen to protect them. Eight weeks of age is the appropriate age to introduce chicks to the older flock. They are completely feathered out and old enough to run and hide or defend themselves against other flock members if necessary at eight weeks.

Chicks should be introduced to older birds through a barrier. A barrier can be an adjacent pen or a metal dog crate. This allows flock members to talk to, investigate, and boss around the younger birds without injuring them. This helps to establish the pecking order without any harmful pecking. Using the playpen method also allows for a faster integration, and after three days, all the birds have usually settled into their role and instructed the chicks on theirs. We always release the chicks by opening the dog crate door at night when all the other birds are roosting to allow the chicks to be free to go about their business in the morning. From that point on, the chicks are part of the flock. We then combine a grower formula into our layer formula, so all members of the flock are getting what they need. At 16 weeks, the feed is switched back to layer formula only. Chicks will become pullets between 8 and 16 weeks. By 16 to 20 weeks, they should start laying, at which point they are full-fledged hens.

What do chicks need? The breakdown on what chicks need in the early weeks is as follows...

Their own separate pen, cage, or bin complete with a heat source until they feather out or are mostly feathered out if kept in a warm environment such as a house or heated barn. Feed either a medicated or non-medicated starter formula. In Canada, medicated starters are not used. They are available in the United States. Vaccines and vitamins usually provided by the hatchery if you get hatchery birds. Not necessary, but if the option presents itself, vaccinating against common chick illnesses is always a good idea. Vitamins can be added to waterers for the first few weeks. Feeder and waterer. A feeder should be easily accessible even for the smallest chick, and waterers should be secure so chicks cannot drown in them. Small bowls will work for this, along with commercial waterers. Safety from the rest of the flock. Chicks should not be integrated into a flock until eight weeks of age. The exception being chicks hatched out by a broody hen who will protect them from the flock.

Medicate or no. Medication is not necessary but a choice that all owners should carefully consider. If a particular vaccine or medication is available for known diseases, a bit of research can help determine if the medication or vaccine is a good idea. Some vaccines are not available in certain areas. It is also a good idea to ask the source where you got your chicks if they have been vaccinated or medicated prior to shipping.

What to watch for. You always want to watch for listless chicks. Chicks will fall asleep quickly when they are only a few days old and often look like they are dead. Startling a sleeping chick should bring it to its feet very quickly. A chick that is listless and does not jump up when startled should be watched carefully and if necessary, separated from the other chicks that can crush him or her. Ensure that any listless chicks are eating and drinking well and are being kept warm enough. Even a small chill from lack of heat source can cause a chick to go downhill fast. Noisy chicks that are huddled together are cold. Tired chicks that are off as far from the heat source as possible are too warm. You should always look for quiet chicks that are spread out around the bin. This is a sign that everyone is warm enough and happy to explore eat and drink without issue.

It is important as well to keep an eye out for crusty bums on chicks. Pasty butt as its often call is when fecal matter accumulates on the chick's vent preventing them from passing waste which can kill them. If you find a chick with pasty butt do not try to pull off the paste. Simply soak the chicks bottom in warm water to help the waste soften and dissipate or soften enough that you can safely remove it without harm to the chick. Remember a loud chick is in distress so if your attempts produce a loud chirpy chick return them to the soak a bit longer.

What to feed your chicks. Chicks should always be fed a chick starter until 8 weeks of age. From 8 weeks to 16 weeks, they can use either starter or grower formulas but not a full layer formula. Older chicks can eat a layer formula that has been mixed with a starter or growth formula. Chicken scratch is not adequate food for chickens of any age as it is a treat only.

## Chapter 7 Get started with Hens

If you consider laying hens instead, then you have managed to escape all the extra work with chicks, and you won't have to wait as long for eggs.

For laying hens, you will need a secure predator-proof coop. This can be a pen in the barn, a chicken tractor, moveable chicken coop, or a permeant building such as a hand-built coop or transformed shed. The coop should have adequate ventilation, bedding, and at least one nesting box per 4 hens.

You have the choice of buying or building your chicken coop. Some instances, buying might be the better option, while under other circumstances building the coop might be the best deal. Either way, the coop should be well ventilated, secure, and house the number of chickens you expect to buy.

Once you have the coop sorted, you will need Feeders and waterers. These come in different styles and models. All have been designed with chickens in mind, so ideally, you are looking for what works best for you. Take into consideration how many birds you have, if you want several feeding stations. Waterers should have a dish or raised surface under them to prevent too much water soaking out into. Feeders should allow birds to feed without getting into the feed and scratching in it. More food is wasted by than anything else. Also, remember you do not want a waterer too big that birds can get caught up in it or younger birds can drown. A contained waterer is always safest. We use top closing waterers placed inside a horse feed bowl to prevent spillage and make it easier for the humans to fill. Our feeder is a top load-free feeder with 12 openings for birds to peck. Over the years, we have tried many different types of feeders and waterers and these were the ones we found worked best for us. You may have to run a few trials before you find the one that works for you and your birds.

When selecting your birds, you want to consider the age of the bird and where you are purchasing the bird from. It is not uncommon for people to sell off flocks of birds that are less than a year old every fall in our local area. These are laying hens in their prime that owners do not want to winter over. As they are under two years of age, it is usually a very good deal to get them in the fall already laying. Beware, however, of anyone selling birds over two years of age as these birds seldom, if ever, lay. A bird laying cycle slows down immensely after two years. When checking older birds, look for fading. Legs and feet, wattles and comb that are all a lighter color are a tell-tale sign of an older bird. Young birds have bright yellow or dark grey and black legs. Wattles and combs are bright red, and the beak is also a bright yellow. As birds age, these areas fade in color, and a combination of fading in these sights shows an older bird.

Quarantining new birds

New birds should always be quarantined for a minimum of two weeks prior to introducing them to a new flock. Ideally, 30 days is best but not always possible. This gives the bird a chance to settle down and get used to you. It will give you a chance to observe the bird for any signs of illness, discomfort, or poor habits like egg eating. Once you are certain your new hen(s) are in good health and okay to add to the flock, then it is time for introductions. Be sure to introduce your bird to the new flock do not just add them during the night or toss them in during the day. It is the best way to start a fight, and the new bird can be severely injured or even killed by other flock members. The best way to introduce new birds to a flock is through play penning. Ideally, a see-through neighboring pen or a large metal dog crate in the coop work best. Leave your new hen in the "playpen" for a minimum of three days. During this time, the other birds can approach, smell, look at and speak to the new hen without hurting her. This allows the established flock to enforce their pecking order while accepting the new bird. If everything looks good and calm after three days, open the door on the crate the evening of the third day. This allows a bit of intro time before roosting, and the new hen should have little to no issues in the morning. If there are still issues, start over for another three days. After that, it is the instigator that gets crated, not the new hen.

By using this method, you ensure that your new hen and the flock accept each other in the least threatening manner.

### *Chapter 8 What not to do with Birds*

- Do not put chicks out in cooler weather until they have feathered out. Once chicks have reached the age of roughly 6 to 8weeks it is safe to integrate them with older birds and to permit them to be outside. Prior to this date they may not be able to maintain their body temperature even during summer months.

- Do not put chicks in with older birds. Older birds will attack chicks unless a mother hen is present to protect them and even then, some birds will try. 8 weeks or older is the recommended age for integrating chicks into the larger flock.

- Do not add birds without introduction or quarantine. By keeping your new birds separated from the main flock for at least two weeks (30 days is preferable) you can reduce the risk of spreading diseases that your new birds may have and be silent carriers. Introducing birds through methods such as play penning (using a dog crate or smaller cage within the coop) allows birds to see, hear, smell and speak to each other without anyone getting hurt. This allows your newer birds to establish themselves in the pecking order in a safe manner.

- Do not put a water that birds can drown in with chicks. This one is fairly self-explanatory, no large open bowls or feeders with water. chicks will drown.

- Do not overcrowd your birds. Birds need an average of 2sq ft sleeping space and 4sqft living space in a coop or pen. Overcrowding increases the risk of disease and heat stroke.

- Do not use cedar shavings for bedding. Cedar gives off a toxic fume that is deadly to chickens. Pine shavings are ideal for chickens.

- Do not use straw or hay for bedding. Straw has an exceedingly small surface layer which makes it harder for it to absorb moisture. Some straw is even hallowed out which encourages mites, fleas and other bothersome insects. Other options such as sand and shavings are much more absorbent. Pine shavings are idea for moisture and for pest control.

- Do not worm your birds unless necessary. Over worming can lead to problems with wormer inefficiency.

- Do not use food grade DE with your birds. This is a long and complicated explanation. Suffice to say DE even food grade DE is hard on birds' lungs (it's like tiny sharp pieces of glass in the lung tissue). It is ineffective on bug control and has no place in or around animals and is not vet recommended.

## Chapter 9 Cleaning up after birds

When it comes to bedding in the coop, you want highly absorbent, something that will not accumulate a lot of bugs or rodents. Your bedding should also be easy to clean up no matter how long it is left and should be highly compostable. Of all the choices that are out their hay, straw, shavings, sand, etc., two stand out as excellent bedding, while two others stand out for the poorest bedding.

Pine shavings are probably the best choice and one of the most popular. Shavings have a wide surface and are more absorbent than other alternatives such as straw, which is narrow and hallow, prone to molding, and hay, a dust source, has zero absorption and needs frequent, if not daily cleaning. The other popular choice is sand. Sand keeps coops cooler in the summer months and helps generate heat in the winter months. It can be scooped out with a hay rack like a cat litter box and does not get muddy, moldy, or dusty. Sands two downsides are that it cannot be used for the deep litter method, and it is prone to attract more bugs like fleas than other bedding choices.

The deep litter method. Here we use pine shavings and the deep litter method. The deep litter method starts with a cleaned coop disinfected with a biological disinfectant that is safe for birds. We use orange vinegar to disinfect and clean our coop, although other lime and commercial cleaners can be used. Once the coop has been cleaned, we put down two large shavings bags, so the coop has a deep bedding layer. Every week we flip and fluff the bedding. In the summer months, we add additional shavings on top of the existing shavings once every two weeks while fluffing and flipping the existing shavings the weeks in-between. We add a new layer (1 bag) every week and continually flip and fluff the bedding in the winter. What this does is allow the bedding underneath to compost. Using this composting method, we are keeping fresh bedding while adding heat to the coop for the winter months. It also speeds up the composting process for the bedding and the chicken poop. Shavings need nitrogen to decompose, and chicken poop is overloaded with it. The result is an extensive clean-out every four months effectively only four times a year. We end up with some excellent compost, and the birds have a dry clean, and warm bedding layer in their coop. The deep litter method works well in northern climates in the winter when it is harder to clear coop waste, and additional heat is needed in the coop. The addition of composting litter for the birds further reduces the need for an external heat source for the birds as the heat from the composting bedding rises through the coops reaching the birds on the roosts.

Should you choose to use another method of bedding, chose your bedding choice carefully. The type of bedding determines how much work you will be doing. Straw and hay should be replaced a minimum of once a week to prevent mold and dust from building up in the coop. If they end up wet, they should be changed sooner. Sand like the deep litter method needs only to be scooped regularly and additional sand added as needed to keep and maintain levels within the coop. The deep litter method can be changed every four months or sooner if there is a lot of moisture in the coop. We change and flip a bit more frequently in the summer months as the extra heat is not needed in the coop during those months.

Disinfecting a coop should never be done with harsh chemicals, as residue can kill your chickens. Bleach is only recommended when cleaning out after a known contagious disease has been present, and the bleach needs to be diluted and rinsed out well after the fact. Lime under the bedding can help reduce pests and the frequent need for cleaning and disinfecting. Vinegar works great and is safe to use at any time for big or small disinfecting. Food grade Diamataeus Earth is not recommended for disinfecting or cleaning, or dusting chickens. It consists of a powder that is made up of microscopic shards. This works well on an insect's exoskeleton, but it can be nasty in your bird's lungs and can potentially kill them if inhaled. Many sites recommend DE, but veterinarians are firmly against using it for birds and other animals.

Fly tape and insect catching devices can be used in the chicken coop so long as they are not within reach of the birds. There is nothing worse than trying to remove the fly tape from an angry hen. Herbs and plants like mint and lavender are not recommended for the coop or the nesting boxes. This has been a widespread practice over the years, but there is zero evidence that herbs of any kind help in any way in the coop. In some instances, adding herbs to the nesting boxes has been known to anger the hens, who then go looking elsewhere to lay their eggs.

Feeding and watering chickens in the coop itself is something one wants to avoid if its possible to feed and water outside. Having food in the coop attracts rodents like mice and rats who take a particular interest in not only the feed but also the eggs. Water in the coop in the wintertime should be monitored to ensure there is zero spillage. The more water and wetness in the coop the higher the chance for frost bite on combs and wattles in the winter weather. Leaving water and feed in the coop overnight is also completely unnecessary. Once birds are on the roosts for the night, they do not come down to eat or drink as they are blind in the dark. Removing the food and water reduces moisture levels and the potential attraction of pests. It also gives you a chance to thaw frozen containers and to wash out feeders, as necessary.

## *Chapter 10 Winter Basics*

Do chickens need heat in the winter? Yes and no. Chickens do need to keep warm during winter. They do this by going up on top of the roosts fluffing up their feathers and wings to trap warm air, and then sitting on their feet. Often, they will also tuck their head under a wing. Hens and roosters alike will again huddle close together on the roosts. The addition of rising warm air and body temperatures helps to keep birds warm. The addition of a vent shaft or strip above their heads helps to draw the warm air and moisture up, preventing frostbite from combs and wattles. A good size coop for the number of birds (not too large and not too small) helps regulate the temperature. The addition of composting deep litter can also raise the temperature of the air several degrees. Ideally, you want birds that do well in colder climates as well. Heat lamps are unnecessary and are pretty dangerous in a coop or barn setting where dust, feathers, and stirred-up bedding can contribute to a barn fire. We have had birds in an uninsulated, unheated winter coop in temperatures as low as -49C, and they did simply fine. So, heat lamps are a risk that should not be considered. Sweaters are also a big no-no. Sure they are cute, but they prevent birds from fluffing up to trap warmth around them. Sweaters make birds colder.

Do Chickens need light in the winter? Again, this is a yes and no question. Yes, they need light in the daytime hours like anyone else, but extra light is up to the owners. Chickens have a natural cycle that sees them slow down in the fall and usually stop laying over winter. The cycle is designed to allow hens to rest and rejuvenate. It also coincides with natures cycle as days are shorter and darker. Hens need between 8 and 14 hours of natural or artificial light in order to maintain their laying cycle. Once the winter equinox passes, birds generally start to slowly pick up in laying as the days begin to get longer again. Adding a light source to your coop and extending the daylight hours can keep your hens laying over the winter. Or you can allow them their natural cycle and rest. Some breeds like the

Plymouth barred Rock can often continue laying through winter months with reduced light cycles, although it is not common for most breeds of chickens.

During the winter months, chickens can benefit from time outside in the sun despite the snow. Some chickens enjoy the snow and will play in it. Its recommended that you clear an area of your yard or pen close to the coop door where your birds can sun and gather and not stand in deep snow. The open coop door will also allow those who do not enjoy the snow the opportunity to go back inside should they be so inclined.

One thing you want to avoid in your coop is drafts. This is different from the ventilation you should have up close to the ceiling, which is above the bird's heads. Drafts at a lower level can prevent birds from warming themselves. Any and all drafts in the coop at the bird's level or below should be patched upon discovery to prevent frostbite and discourage predators from entering.

## *Chapter 11 Dealing with Predators*

Chickens are tasty. Everyone knows chickens are tasty; even chickens know that chickens are tasty. So, it comes as no surprise that there are many critters skulking about looking for a way into your chicken coop and chicken pen.

Chickens main predators are hawks, foxes, coyotes, minks, raccoons, and in some instance's possums. All these predators prevent certain challenges for protecting your birds. All of them approach in different manners as well. So, we will break it down by attack style and predator.

Air- An attack from overhead is always a bird of prey, usually a hawk or an owl. These birds will swoop down from a tree, barn roof of direct from the air. They attack, then pluck and eat at the scene. On occasion, if the bird they catch is small enough or if they have young in a next, they will attempt to carry off their prey. Air attacks can be prevented by using several methods.

Cover the exterior pen or chicken run with Fencing, fishing line, tarps, nets, etc. If the cover cannot be passed, they will not attempt to attack. When using fishing line, additional cds or dvds that are reflective and shiny can be tied to the lines to discourage overhead strikes further. A fake owl moved regularly can also discourage hawks and territorial birds or prey. A larger animal on the ground also helps. Hawks and owls are defenseless on the ground so a large goose, dog, goat, or another animal that could be a threat will also prevent birds of prey from swooping down. Do not use a pig though as they will kill chickens for fun.

Diggers, jumpers and chasers, these are your coyotes, foxes and neighborhood dogs. Proper Fencing using the Fencing out method as well as a top cover can prevent a direct attack. Putting Fencing under the ground1 to 2 ft and out 1 to 2ft from the fence line can stop the digging as well.

The sneaky bandits… these are your coons, possums, and minks. These animals gain access to your coop or run by going through weak or open holed Fencing they can rip tear or squeeze through (NEVER USE CHICKEN WIRE). They also find weaknesses in coop walls and floors. The best way to prevent them from entering is sealing up all cracks and holes no matter how small. Coons can also open various types of hooks and latches so make sure your lock on your coop is one that cannot be opened by an animal.

## Chapter 12 Know before a Crow

Some chickens are sex-linked breeds, such as the Cream crested legbar and red and black sex links. Sex link birds means the breed has different coloration and characteristics at birth that differ from the opposing gender. Boys are one color, girls another. Sex link birds are great as you do not run the risk of getting a rooster if you do not want or cannot have one. Other breeds have different characteristics that can be monitored as they grow to determine gender. Some of those characteristics are…

Darker comb and wattles than other birds its age.

Feathering out slower than other chicks the same age

Thicker legs than other chicks its age

In some birds like the Plymouth barred rock,

newborn chicks have markings that give away their gender. For example, Plymouth barred rocks will have a yellow spot on the top of their head for the first week. A large circle indicates a female, while a misshapen elongated circle indicates a male. The markings will eventually disappear..

Lighter or darker color than other chicks the same age (again, barred rocks have this. Males are lighter in color than the females)

Neck and saddle feathers. As birds mature and feather out neck feathers will be arrowed sharp (male) or rounded (female) saddle feathers located just above the tail on where the back, and the tail meet will have wither elongated arrow tipped feathers (male) or rounded thick feathers (female).

The most notable sign however is crowing which can start as early as 3 months but is much more prevalent and stronger at 5 to 7 months of age.

If you watch for these signs and you have birds that exhibit several chances are you have a roo. Young roos are easier to rehome than older ones so the sooner you notice these signs the sooner you should tag that bird for observation and if necessary, rehoming

## Chapter 13 Training your chickens

Yes, you can train your chickens. Chickens are smart enough to learn their name and often try to chat with you if you call them by their name. They can be taught to come when called (especially useful when putting them back in the coop at night) and even taught to recognize signs and symbols. My favorite trick is getting a two-year-old blue Orpington named Emma to beg and jump up for a treat. Chickens are the new dog.

Training your chickens consists of repeatedly offering treats for behavior or by responding to a call. We trained our birds to come using this method.

With your birds in an enclosed area, drop treats on the floor (table scraps, mealworms, scratch, etc) and call them. We always yelled goose goose goose when we trained the geese; the chickens caught on and started coming too.

After several days of calling and dropping treats in an enclosed area, move a bit further out. Whether it's the door of the coop or out in the pen, move away a bit from your usual spot and drop the treats while doing your call.

Gradually move further and further away, calling louder and louder. After a period, your birds will learn to come to you for the call no matter how far away you are. Just remember always to offer a treat when they come.

Using the same method of offering treats your birds can be trained to jump up for treats, to come for treats, to sit patiently and wait for treats, etc.... chickens are smart enough that they can teach the family dog a trick or two.

## *Chapter 14 Daily Task List*

When taking care of your chickens, there are certain things you should always do daily.

Feed and water your birds in their regular spot. Stop to observe who is eating who is waiting to eat. This should give you an idea of who the alpha chicken is, but also which birds to check for weight. If the waiting birds seem underweight, be sure to add more feed when their turn comes.

Check your waterer for growth of mold and algae. Some forms are highly deadly to birds.

Check your doors, latches, and Fencing for signs of an attempted attack or potential source or entry for an attack. Repair anything you find right away.

Check nesting boxes for signs of egg eating, parasites, or excessive waste that would indicate birds are roosting in the boxes. Clean disinfect, close off boxes at night or move birds once lights are out to teach them to roost elsewhere.

Look for overbreeding signs (missing feathers neck

and shoulder areas) or molting (missing feathers all over).

Look for signs of worms or disease in chicken poop. Worm all birds immediately and try to find any sick hens.

Look for hens that stay away from the flock (not the nesting birds) in corners by Fencing under trees etc. These birds may seem particularly quiet and listless. Check for signs of illness or injury and quarantine immediately (having dog crates on hand are great for this)

Check for fly levels and any mites' fleas or bugs on birds particularly under their wings.

Watch your flock for signs of stress, happiness etc. Knowing your bird's regular behavior by watching them for a few minutes throughout the day every day will help with this and can potentially prevent a major issue later on.

Check your roosts and bedding to see when and if cleaning is necessary.

## *Chapter 15 Chicken diseases*

Chickens are notorious for hiding illness and injury. It is commonly believed they do this to make them less of a target from predators and members of their own flock who will attack them to protect the rest of the flock. A sick bird is a major attractant for a predator, potentially putting everyone at risk. In most instances, by the time you notice an illness in a chicken, it's too late to do anything about it. On the rare occasion, if you catch it early enough, you may be able to save your hen.

**Ten common Chicken diseases**

1. **Fowl Pox**- easily recognizable by white spots and patches on the comb mouth and wattles. It will also produce ulcers in the mouth. A vet can treat this, and the bird should be quarantined as it is contagious. Left untreated, the mouth ulcers can prevent a bird from eating. Soft food is recommended during treatment, so either fermented chicken feed or soft dog or cat food.

2. **Fowl Cholera**- spread by other animals; it is a deadly disease characterized by darkened wattles and comb, listlessness, and yellow diarrhea. It attacks birds over five months of age, and there is no treatment. Birds should be quarantined immediately, and bodies properly disposed of.

3. **Infectious Bronchitis**- highly contagious. This one is airborne and spreads quickly. Characterized by coughing and listlessness, there is no treatment. Birds infected should be quarantined immediately.

4. **Botulism**- Tremors and falling feathers are characteristics of this disease. It is lethal, but some birds can recover. Treatment is available through your veterinarian.

5. **Infectious Coryza**- swollen heads and eyes, nasal discharge are main characteristics. This is a lethal disease, and birds should be removed from flock immediately. No treatment

6. **Thrush**- spread from contaminated water, characterized by discharge from the crop and a messy backside. Treatable with antifungal meds from the vet.

7. **Newcastle Disease**- almost always deadly in chicks older birds usually survive but are carriers. Characterized by discharge from eyes, mouth, and nose,

respiratory issues. Vaccine is available.

8. **Pullorum**- characterized by comprised airway in older birds, no signs in chicks. Deadly and infected birds should be dispatched immediately to prevent spread. No treatment.

9. **Avian Influenza**- inflamed face, severe diarrhea, red patches on legs and face. Birds must be dispatched immediately to prevent spread. Highly contagious, no treatment. Can infect humans.

10. **Bumblefoot**- when birds get infection inside cuts on their feet. Treatable by vet's infection is removed from foot and bird is quarantined to stay off the infected foot. Quite common in free range chickens.

*Epilogue*

It was estimated that there were 23.5 billion chickens on the planet in 2018. Estimates are running about 50 billion in 2020. That is roughly seven chickens per person on the planet. These amazing birds have their own language, alarm call, and courting behaviors. They can be trained to do tricks and can see movement 20x faster than humans can. A chicken's brain even works independently with each eye. A chicken can watch its food and its surroundings for predators at the same time. It comes as little surprise that they are quickly becoming the pet of choice for urban and rural homesteaders. Back yard coops are popping up all over, and people of all ages are now learning just how amazing these little birds are. By adding backyard chickens to your home, you are adding not only a valuable food source, but an amusing pet and a bug-eating dinosaur all at once. Hopefully, the information we have shared in this book will help you to better navigate chicken ownership, but there is one last lesson to be learned.

Chicken Math

2 chickens = 2 Chickens

4 chickens =2 chickens

8 Chickens= 2 chickens

10 plus Chickens = 2 chickens

20 plus Chickens = needs more chickens

If you have enjoyed this book, please return to Amazon, and rate and review it. On the following pages you will find an excerpt from my new book "The Naked Ugly Truth About Homesteading and Self Sufficiency" available now on Amazon.

## Preface: The Ugly Truth

There are so many books, and blogs, and YouTube channels of people homesteading. Pages and pages of articles and videos of things to do and to try. These are the pages and articles that promote the pretty side of homesteading.

If you believed every homesteader out there promoting their page or book, you would be inclined to think that homesteading is the magical solution to everything. The escape from the city to the idyllic little country home on acres of rolling hills and pastures with a big, beautiful barn in the background. Chickens running everywhere. These are the books and internet pages that make it look so easy, and so ideal, and so affordable that you cannot help but dream of your perfect little homesteading life every chance you get.

You plan out your garden and the different crops you will grow. You add a kitchen garden to have everything you need to prepare meals from scratch and prep and preserve all the food you will grow. You research all the different types of chickens and ducks, cows and horses, goats and sheep. Then you hit the realty pages looking for that perfect little acreage where you will grow and provide everything your family needs, commercial farms, and industry be damned.

Well, in a perfect world, those dreams would be not only attainable but easily sustainable, and they would go swimmingly along without problems or issues of any kind. We do not live in a perfect world. Many books and internet gurus do not paint an accurate picture of what homesteading is and is not. They do not explain exactly what goes into running a homestead, being self-sufficient, or living off-grid.

They show the result of hours of work, preparing, and planning to film that video or take that photo. They may be showing a bumper crop of cucumbers but not showing the endless rows of failed tomatoes. They may tell you all the stuff they preserved but not disclose how many jars did not seal or how much food was wasted. Very few individuals promoting their books and videos will give the cold hard truth about homesteading and self-sufficiency.

Primarily because that is not what people want to hear but also because they would not make any money if they did. In all honesty, if you want your homestead to survive and succeed, then you need to know the truth about homesteading and self-sufficiency ahead of time, and it is not what is often portrayed.

This book is designed to lay it all out. The good, the bad, and the smelly. What is homesteading really like? Can you attain Self – Sufficiency or is sustainability enough. There is no mincing of words here, no perfect tomatoes but lots of chicken poop. So, dive in, have a good read, and then… go create your ideal homestead.

Chapter 1: Homesteading is Hard.

Homesteading is hard. That is one point that needs to be made clear from the very beginning. I'd scream it from the barn roof if I thought it would make a difference. No matter what kind of homesteading you do, it will be hard. There are endless chores and tasks that will have to be done daily, weekly, monthly, and yearly. As a homesteader, you will need to organize, prioritized, and complete those tasks in a timely manner. If you don't, crops fail, animals die, and your homestead will be a source of endless chaos. Many of the chores on a homestead will have to be done regardless of illness, injury, time constraints, and financial status. Animals do not feed themselves; gardens do not water themselves; weeds do not weed themselves, and food does not store itself.

Regardless of what else you must do or where you must be, or how you feel, you simply must do these tasks, no excuses. You do not want your dreams to fizzle.

Homesteading is not cheap, and it is not easy to make a living off it. There are few individuals who homestead that do not have some sort of outside income. Even those on YouTube are garnering money from ads and sponsors for their videos. Many homesteaders do video vlogs, online websites, and blogs, write books, teach classes, and give tours to supplement their income. Many homesteads sell market garden baskets and veg and egg shares. Some raise and sell animals.

All of this involves extra work for the extra income. Most homesteaders have regular day-to-day jobs off the homestead, which reduces hours they can spend on their homestead. It increases the amount of work that needs to be done when they get back home.

In some instances, both partners will be working off the homestead, or one will stay home and the other work. The expectation being that the partner who stays home does the bulk of the work and sees to it that the homestead produces food and crops to offset the income lost.

This is quite different from the painted picture, where you see large families happily homeschooling and homesteading. The reality of those videos is there is often an extra income earner, plus the revenue from the videos, many hands to help, and many mouths to feed. Older children are usually relatively independent and can navigate chores and work assignments with little supervision. They often help younger siblings. This is not the case for a starter homestead with just two adults or two adults with two exceedingly small children.

When we started our previous homestead, our girls were still in diapers and could only carry empty buckets to the barn. They spilled many cups of feed, broke countless eggs, stepped on many chickens. They even had their hair chewed on by a goat, which they seemed to think was terribly funny. Fast forward to our current homestead, and our girls can haul water, feed the animals, help with seeding and harvesting, and even beekeeping, but it took a while to get to this point, and for that time, it was just my husband and myself building a dream.

We supplemented our income with market garden shares, animal breeding and sales, egg shares, and extra income from his job. We share the workload with me, taking the daily chores and lighter work and him taking on the bigger, heavier tasks in the evening or during weekend hours.

Homesteading is not the solution to all the world's problems. Okay, yes, we would all like to know where our food comes from and reduce greenhouse gases and carbon footprints. Homesteading is by far a step in the right direction, and self-sustainability is a definite goal to aspire to. Still, the reality is that homesteading must be done right to significantly impact the environment.

Some homesteaders are doing it all wrong and are increasing their carbon footprints instead of reducing them.

This is one of the main reasons why it is essential to know all the shortcomings of homesteading, self-sufficiency, and sustainability before embarking on that wild dream. Knowing what it is really like can help avoid many hidden pitfalls and lower some of the higher than reasonable expectations that come with the homesteading dream.

Things like:

- How much of your food can you reasonably grow on your homestead?
- How much land do you need to be entirely or even partially self-sufficient?
- Do you need to go off-grid? What is it like?
- What animals can you have on your land?
- How much land do you need to raise your crops and animals of choice?
- What does it cost for land, animals, seed, tools,

equipment, feed, and supplies?
- How much work is involved?
- What will you have to give up?
- Is the gain worth the sacrifice?
- Is everyone on board?
- What if SHTF?
- Are you prepared to face nature head-on?

These are just some of the questions that every homesteader should ask before they find themselves knee-deep in composted chicken poo. This is the shortlist. Being a homesteader means being prepared for as many outcomes as possible and as many trials as possible.

To be a homesteader, you need the skills of an engineer, a carpenter, a chef, a farmer, a vet, a mathematician, an inventor, a butcher, and in some cases, even a marksman.

To run a homestead home, you need to know how to cook and sew and bake and preserve. Knitting and crocheting, washing and mending, budgeting, and organizing.

These are not all skills everyone has upfront, but they are skills that every homesteader should be prepared and willing to learn. You never know when you must get creative to keep an animal contained or a predator out. You can never guarantee an animal will birth without complications or will not get sick or injured. If you grow and preserve food, you will have to develop creative ways to serve and store it.

There is so much to learn about homesteading that you could spend an entire lifetime doing it. In fact, that is precisely what homesteaders do. Make no mistake, you will never stop learning, but preparation is vital, and you cannot prepare if you spend all your focus on the pretty side of homesteading.

## Chapter 2: Self Sufficiency

If I had a nickel for every book or article, or video I saw detailing how you can be self-sufficient on an acre of land, I would be so bloody rich. It's just not true. Not at all.

That said, you can have a level of self-sufficiency on an acre or even half an acre or even a third of an acre, but you cannot be "self-sufficient."

To be even partially self-sufficient, you need to provide for the majority of your family and farm needs.

- All your food, including legumes, vegetables, fruits, grains, and proteins, will need to be produced. An ample enough garden space or several spaces with good quality soil is essential for this. - If you need wood as a source of heat, you will need to have a woodlot on your property to supply that resource.

- Depending on whether you are on or off the grid, you will need a water source and water storage. A water source can be a well, public line, or a body of water. In the event of a well or body of water, you will have to refrain from putting livestock, gardens, or other such things on, in, or near these to prevent contamination or damage. Water storage will also require a reasonably good-sized area to store sufficient water for cooking, bathing, watering of crops and livestock.

- You need a source of income to pay for things like healthcare, medicine, gas for your car, equipment, and tools.

- If you plan to keep bees, you will need sufficient space that the bees are not in constant contact with humans or livestock but properly set up to be south facing and secure from high winds.

- If your protein comes from animals or animal by-products, then you will have to have sufficient space for those animals. You will also need room to grow and store feed for those animals. You will need to provide shelter and protection for those animals.

- You will need a power source. If you plan to be off grid, you will need to have space to put wind and solar power equipment. If you stay on the grid, you will need to have a backup resource such as a generator or wood for a power and heat source in the event of a power outage.

These are just some examples; the reality is that it is darn near if not completely impossible to be self-sufficient on any amount of land. Sustainable and partially self-sufficient yes, but completely self-sufficient, no. You cannot provide your own health insurance or money. Few people know how to forge metal or have the resources to create it.

There will always be a need for some equipment that will have to be bought, bartered, or traded. For that you need transportation and gas. Even free items still need to be picked up and transported.

That said, there is a very real prospect of attaining a level of sustainability on your land. The question is, how much? What depends on the land and what you want to do with it.

For example: If you want to raise all your vegetables and some protein such as chickens and rabbits. This can easily be done on 2 acres or less. If you want a milk source such as a goat and you want to be able to feed that goat from your own land than you will need at least three to five acres.

If you need a wood source, but do not have sufficient land for a woodlot, then you can outsource the wood and concentrate on food.

The key is planning and organization. A small parcel of land can still produce a large amount of food. Chickens and rabbits take up next to nothing for space. Even a goat can be sustained on a small parcel of land if you plan to outsource the feed.

Attaining a level of self-sufficiency and sustainability is a very realistic goal even on a small homestead. Expecting to be completely self-sufficient is just not realistic. We will always need other people for some skills we do not have, and to produce products we cannot make. The world is designed that way and has been for some time.

Much like the happy visions of perfect homesteads, self-sufficiency is extremely hard to achieve, if at all.

Ideally, when you talk of self-sufficiency, real self-sufficiency, you are saying your homestead is "needing no outside help in satisfying one's basic needs, especially with regard to food production." (Myriam dictionary). No outside help is kind of impossible when you think about it. Even Fencing must come from somewhere.

So, what is possible as far as self-sufficiency is concerned. Well, a lot, if it is well planned and well executed. It depends on your goal.

A "self-sustaining" homestead (able to continue in a healthy state without outside assistance, Myriam dictionary), can be achieved if you plan to only grow vegetables and fruits, for example. Once you seed you can save your seeds, season after season, and not require any additional outside sources for seed.

A "sustainable" homestead (able to be maintained at a certain rate or level, conserving and ecological balance by avoiding depletion of natural resources, able to be upheld or defended. Myriam Dictionary) is probably the most logical and easily attainable homestead and can be achieved by careful planning, sufficient land, and proper execution. I.e.. You do not bite off more than you can chew and plan for the future as well as the here and now.

Here we run a sustainable homestead. We have rabbits, chickens, ducks, goats, and bees. We keep four barn cats for rodent and pest control, and two dogs for predator control and companionship. We have three massive gardens and four smaller ones in addition to a large, raised pumpkin patch, an orchard and food forest.

We use a rain catchment system, and we outsource our wood.

We also outsource a pig each year from another local homesteader and we sell our veg and eggs to friends and coworkers.

We buy our feed from the local feed store but supplement our animals with veg from the gardens and food scraps from the kitchens. If push came to shove, we could feed our own animals for a period without outside assistance, but this is carefully planned and to date has not been needed.

Our homestead is designed with the future in mind, and we have laid out all our planting based on future plans and needs. There is room to expand on projects we have not gotten to yet and plenty of space for additional projects we may not have thought of yet. To do this, we had a very clear set of goals in place before we purchased the land and a checklist of what we absolutely needed to have and what we could do without.

Our house is designed in such a way that we can continue to go about our daily routines without a power source. That said we still require a generator to save our freezers from failing but we don't put all our eggs in one basket and use several methods of food preservation to keep our stockpile from spoilage.

Should there be a power outage such as the one we had last winter where power was out to much of the province for over 5 days, we have a heat source, back up compostable toilet, cold storage for food and ample sources of light. Our car serves as a backup charger for things such as cell phones and we keep plenty of gas on hand.

We can sustain our home for long periods of time without the assistance from the outside world if needs be. That said we still use resources from off our homestead to build up feed and food storage, provide electricity and a variety of personal and household goods. We do however carefully plan these purchases with several things in mind."

1) budget homesteading is not cheap and money resources need to be used according to priority and necessity.

2) the impact on the environment, one time use items are a no go here. Paper towels went out the window years ago and even toys for the girls are

carefully considered for their long-term playability and interest level; and

3) is it beneficial. There is a big difference between wants and needs.

When you have a homestead there is a long list of needs so there should be a short list of wants.

Considering workloads, space, environment, and budget there are few "extras" that come into our home that are frivolous. Even our animals must serve at least two purposes, or we do not have them.

- Our chickens provide meat, eggs, and compost as well as bug control, as do our ducks.
- Rabbits are meat, compost, and pelts.
- Dogs are predator control and companionship,
- Cats are big time pest control for the barn and the house.
- Goats are milk, meat, and compost.

By ensuring that the things we grow and raise here are dual purpose we increase the sustainable factor for our homestead. By ensuring that our house is not filled with useless stuff we have more time to appreciate our home and our homestead and to focus our attention where it's needed. Making it as sustainable as possible.

I cannot see homesteading any other way. But the premise of homesteading has not always been that way. Its initial purpose was to get people to move to the western regions of Canada and the US to clear land and expand territory.

Sustainability was not on their minds in the way we see it today. In the initial days of homesteading sustainability was necessary for survival. Self - Sufficiency was something they strived hard for but even then, settlers were dependent on general stores for staples and equipment, neighbors for help with difficult tasks and harvests and trade, veterinarians when necessary etc.

The original homesteads were nothing compared to today's adventures and to be a true genuine homesteader by their standards you'd have to even give up solar power and composting toilets... it's darkness and an outhouse on that regard.

Often when people think of homesteading, they don't consider the harder aspects of it or the small costs that add up. This book is designed to lay out the initial set up of a homestead and show the downsides as well as the good tips for various types of set ups as well as their costs. From setting up the gardens to the horrors of harvesting we are about to get to the ugly naked truth about homesteading and self-sufficiency.

## Chapter 3: The Homestead

So, what is a homestead? Well, there are many names for it. In some parts of the world, it is called a hobby farm, a croft, a small holding, an allotment, gentleman farming etc. All these terms are used for a form of non-commercial small family farm. The biggest difference between these terms is how the farm is set up. Ideally when most people think of homesteading, they do not think of machinery or large commercial barns. They picture a farmhouse with a small barn a few outbuildings and gardens. Usually there is an orchard as well and there are always chickens in the picture. Not all homesteads have chickens but for some reason when people think of homesteads the chickens are always in the picture.

When I think of homesteads, I usually picture a small farm of 10 acres or less... sometimes 20 but usually never more than that. But there really isn't any set dimension involved in homesteading. The original settlers were granted parcels of 160 acres under the US and Canadian homesteading acts. The trouble with that much acreage is that you need machinery to run it, or a team of dedicated people to help you. The exception is if you are turning out livestock to graze that land but even then, you would need a lot of livestock. Which is why most large parcels over 20 acres are usually considered "farms" since farms are associated with machinery while smaller parcels without machinery are considered homesteads.

Homesteading however is more a lifestyle than an actual industry or profession, hence why machinery is normally not associated with homesteading. It is viewed as the more traditional type of farming.

The ideal homestead will have sufficient land to produce the required amount of food, sustain the required livestock and the required amount of storage to support them. From there priorities can go in many directions. What is good for one homestead may not work for another and they can all be as different as peas and carrots. This is why homesteading is really a lifestyle. Some homesteads are sustainable, some are not. Some prefer permaculture while others go for traditional farming methods. Some are rigidly laid out while others just put things wherever preferring a more eclectic method. Beware the eclectic method.

When the Storey family put out their book on country living years ago, they included a list they had found written by Edward and Carolyn Robinson. It was "the have more" list. This list is ideal for anyone planning a homestead that does not want it screwed up from the start.

- All land should be useable to some degree at some point.
- Gardens should be accessible and have good sun
- Pasture divided and fenced for rotations.
- Herbs, kitchen gardens should be close to the house
- Lawn, shrubbery easily maintained and attractive
- Children's play area easily viewed from house and away from street or screened for privacy if possible/ necessary

- Compost should be located between the barn and the gardens

- Trees should be spaced so as not to crowd

- Barn should be close enough to supervise livestock from house if necessary.

- Adequate storage space for tools equipment, house etc

- Space for good home workshop

- Cold storage for veg and canned goods

- Fencing to allow livestock to be turned loose from the barn

- Space for the home freezer, laundry, and fireplace wood.

- Orchard should not shade the gardens.

Now this is by far not their full list, but this is what I would deem practical for a modern homestead. Not all of these are always possible but things like being able to turn your livestock loose from the barn is a definite luxury.

You do want your barns and coops close especially if you live in the northern hemisphere where there is snow in the winter. The same can be said for beehives. You don't want them too close, but you don't want to be waist deep trying to dig out hives after a storm on the far end of a 5-acre plot of land.

While there are a million scenarios as to what a homestead is and isn't, the ideal homestead is very much a personal affair. Some may want to be as sustainable as possible while others may just want veg and a few small animals, others may want the whole enchilada. A homestead is very much what you make of it. The question is what can and cannot you make of it.

The Naked Ugly Truth About Homesteading & Self Sufficiency

Now Available on Amazon. March 2021

www.earthharmonyliving.com

www.ingramcontent.com/pod-product-compliance
Lightning Source LLC
Chambersburg PA
CBHW070416220526
45466CB00004B/1426